THE ENVIRONMENT

GLOBAL WARMING:

Causes, Effects and Mitigation Measures

VOLUME 2

ISAAC ARIGBEDE

(C) 2016 I.O. Arigbede

All rights reserved. No part of this publication may be reproduced, stored in a retrieval system or transmitted in any form or by any means without the prior permission of the author.

Printed by: AROLIS SERVICES,
 3, Sewage Avenue,
 Behind Jakande Estate,
 Abesan, Ipaja,
 Lagos State.

Phone: 07032550987, 07012987048.

PREFACE

Global Warming is one of the major and most discussed global issues in the world today. Yet, billions of people did not understand this life-threatening global problem. Besides, all forms of vital discussions, wise suggestions and reasonable resolutions at world conferences achieve very little success among the participating nations. Why is it so?

THE ENVIRONMENT Global Warming: causes, effects and mitigation measures volume two is written to enable the readers (the educated, the students, the professionals and the non-professionals) understand global warming and to educate them in the most simplified manner with a view to taking necessary actions that will make the world a safe place to live and prosper.

This handbook has eleven chapters and a total of forty two pages. This book discusses global warming in a simplified manner to enable readers have a sound knowledge of this ever growing problem; the different causes, the effects and the necessary measures required of individual, group, organizations and government with a view to preventing, reducing or controlling the severity of this global problem.

ACKNOWLEDGEMENT

This book has been made possible by the special grace of God Almighty who directed and guided me. I greatly appreciate the priceless gift of good health, precious time for this intensive research work, knowledge and wisdom God gave me in completing this work. I remain ever grateful to Almighty God.

I will also like to express my sincere appreciation to family and friends who contributed greatly to the successful completion of this book.

CONTENTS

	Pages
Preface	iii
Acknowledgement	iv
1. Facts about Natural Global Warming	1 – 5
2. Global Warming	6 – 9
3. Burning of Fossil Fuels	10 – 13
4. Ozone Layer Depleting Compounds	14 – 16
5. Ruminant Animals and Global Warming	17 – 19
6. Termites' Role in Global Warming	20 – 21
7. Paddy Rice Farming and Global Warming	22 – 23
8. Landfill, Wastewater Stagnation and Global Warming	24 – 26
9. Natural Causes of Global Warming	27 – 30
10. Other Natural Causes of Global Warming	31 – 36
11. How Greenhouse Gases Cause Global Warming	37 – 42

CHAPTER 1

Global warming is as old as the creation of our planet. The provision of warmth by nature is as essential as other natural provisions namely air, water, land and cool season to mention just a few. All these work perfectly well together to ensure continuous existence of humans, animals and plants. There was no cause for alarm because no hazards or adverse effects were experienced during these early years of creation of the earth and its inhabitants. This type of global warming is **needed** or **compulsory**. This period can be referred to as the **pre fossil fuels era**.

The discovery of fossil fuels (gas, crude oil and coal), its extraction, its different processing/refining, its numerous uses drastically changed the scenario discussed in paragraph one. In addition to fossil fuels discovery, increase in human population, rapid scientific and technological development, rapid economic growth (industrial revolution era) and many other human activities result in the **unwanted global warming** and this ultimately contribute to climate change.

In this line of reasoning, this book will focus on vital topics and sub topics that will educate, inform, enlightened and/or improve readers' knowledge and understanding about global warming; its causes, its effects and the necessary mitigation measures. But first, what facts do the structure and composition of the planet earth reveal about natural global warming?

FACTS ABOUT NATURAL GLOBAL WARMING

The position, the shape of the earth, the chemical composition of the earth, the physical components of the earth, its rotation and revolution support natural global warming and these make it possible for humans, animals and vegetation to exist. These facts are some of the major reasons why there is a constant natural control or regulation in the intensity of sunlight or heat and the level of coolness experienced on earth by variety of natural events or natural phenomena.

1. Earth's Position

Earth is the third planet away from the sun. The nine planets starting from the closest to the sun are; Mercury, Venus, **Earth**, Mars, Jupiter, Saturn, Uranus, Neptune and Pluto. While the earth remains the only habitable planet, the remaining eight planets are either too hot or too cold for any forms of life to exist. The temperatures of the other eight planets depend on their nearness or their distance from the sun. No matter the rate of alterations or variations in the earth's temperature, the earth still remain the only home for all humans, animals and plants.

2. Shape of the Earth

The earth has an oblate spheroid shape. That is, it has the shape of a sphere but with flattened surfaces at the north and south poles causing the middle to bulge. This middle bulge known as the equator, has the longest distance or diameter while the shortest distance is at the north and south poles. The diameter decreases equally with equal movements vertically away from the middle bulge (equator) northwards and southwards. The middle bulge keeps the earth's shape in a balanced position and it ensures perfect rotation or spinning on its axis to enable every region receive sufficient amount of sunlight in turns. This results in natural global warming. This continual process provides daylight and darkness as the case may be for every 24 hours or on a daily basis.

3. Chemical Composition of the Earth

The planet earth is made up of two major chemicals and many minor chemicals in different proportions. These function to support life on earth and to maintain adequate level of warmth or to ensure a bearable natural global warming through series of chemical reactions. The major chemicals are; Nitrogen (N_2) 78.00% and Oxygen (O_2) 20.95%. The minor chemicals include; Argon (Ar_2), Carbon dioxide (CO_2), Water vapor and so on.

4. Atmosphere and Physical Components of the Earth

Starting from the outermost layer, the planet earth is made up of the following layers; the crust, the mantle, the outer core and the inner core.

4a. The Atmosphere:

Before arriving at these four earth layers, the entire earth is covered by layers of gases generally known as the Atmosphere. Exosphere is the outermost or farthest gas and troposphere is the innermost or closest atmospheric gas to the earth's crust. These gas layers perform various functions ranging from absorbing sun rays to reducing and regulating the temperature and the heat on the earth's surface. These keep the earth warm.

4b. The Earth's Crust:

The earth crust is the outermost layer of the earth. It is the thinnest layer. It is a layer consisting of solid rock. It is made up of the continental and the oceanic crusts. The earth crust temperature can rise up to $1000^{\circ}C$. The earth crust temperature increases with depth. The greatest component of the crust is silicon. Other chemical substances include aluminum, iron, calcium, potassium and sodium. All these combine with oxygen to form oxides. The high temperature and the chemical reactions result in the natural warming of the earth.

4c. The Earth's Mantle:

The earth's mantle lies between the hot outer core and the thin crust. It is the thickest layer of the earth. The mantle temperature varies between $1000^{\circ}C$ to $3000^{\circ}C$. The mantle temperature increases with depth. It is rich in magnesium, aluminum and silicon.

4d. The Earth's Outer Core:

The outer core contains a very dense liquid iron and nickel. It also has small amount of silicon and oxygen. It lies between the mantle and the inner core. The temperature of the outer core is above 4000°C.

4e. The Earth's Inner Core:

This is the center of the earth. The inner core also contains iron and nickel. This is solid with iron as its major chemical component. The temperature here is even higher than the outer core, it is 6000°C. The inner core is solid due to its higher pressure than the outer core pressure. In conclusion, the earth through convection receives the required amount of heat despite the very high temperatures of the layers below the earth's crust. This ensures a naturally balanced global warming.

All the physical features and the series of chemical reactions on earth perform very vital regulatory functions to keep the earth at a bearable level of temperature. Therefore, all these natural processes continually ensure a natural global warming on the planet earth.

CHAPTER 2

GLOBAL WARMING

What is global warming?

Global warming is the gradual heating of Earth's surface, oceans and atmosphere. Global warming is also the increase in Earth's surface temperature due to emission of greenhouse gases such as carbon dioxide, methane, nitrous oxide and halocarbons.

Global warming is just one aspect of climate change (long term or permanent change in weather pattern) that has to do with average global temperature rise. Global warming did not just happen one day or by chance. Global warming can be traced to two sources. These two sources are living and non-living (natural) things. These sources are known to be responsible for the weather and climatic variations experienced by human and all other living things on Earth. The different causes under the human and the natural sources, the bad effects on living things and the environment and the practical measures of reducing or controlling the bad effects will be discussed in this section.

CAUSES OF GLOBAL WARMING

The two sources of global warming can be categorized into living and non-living (natural) things on Earth. The living things include human beings, big and small animals.

CONTRIBUTION OF HUMANS TO GLOBAL WARMING

The major human activities which continue to contribute greatly to global warming include but not limited to deforestation, burning of fossil fuels and the use and release of ozone layer depleting gases.

DEFORESTATION

Deforestation is the rampant removal or clearing of trees without replanting to re-establish the forest or maintain its existence. Also deforestation is the permanent destruction of forest in order to make it available for other uses. Forest is made of different trees and other green plants. But trees are the backbone of forests. Some of the trees include Kapok, Mahogany, Cedar, Opepe, Cherry, Tick, Oak, Afara, Pine, Iroko, Kerala, Gujarat and Willow to mention just a few.

How Deforestation Causes Global Warming:

When there is deforestation, carbon stored in the trees is released to the air in the first instance. Also, carbon dioxide that should be used by forest trees will remain in the air. The result of the unused carbon dioxide in the air is greenhouse effect and this will finally cause global warming. What are the reasons for deforestation?

Reasons/Causes of Deforestation:

Some of the reasons/causes of deforestation include population growth and expansion, war, oil exploration, infrastructure development, mining, poverty, cattle ranching, rampant logging of trees for home, for export for industrial uses, for agriculture/farm establishment, for industries. Others are; forest fires, expansion of water ways, lack of or unstable forestry policy and so on.

Effects of Deforestation:

Deforestation has many adverse effects on humans, animals and soil/land. These include but not limited to soil erosion, climate imbalance, floods, global warming, death or extinction of wildlife and their habitat, loss of jobs, scarcity of forest produce, loss of topsoil, reduced revenue generation, desertification, disruption of the forest ecosystem and so on.

Prevention of Deforestation:

Some practical actions that can be taken include using recycled items, selective felling of trees, planting trees for replacement, using extension workers for forest advocacy, paying foresters good salary, establishing parks, supporting environmental awareness organizations, practicing crop rotation instead of land rotation, government should embark on forest development project, treating or cutting diseased trees promptly and creating jobs for inhabitants around forest.

Advantages/Benefits of Forestry:

Forest influences weather and climate condition positively, it reduces global warming, prevents flooding, contributes to state and national revenue, serves as a wind breaker, provides variety of fruits, nuts and seeds.

Furthermore, forestry is a reliable source of herbs (e.g. cacao and red cedar trees), it is a source of industrial materials, it is a wildlife habitat, it provides job opportunities, it helps water quality, keeps the air clean, reduces wildfires, it is an ideal environment for research, it is a good venue for recreation, it enriches the soil with green manure, etc.

CHAPTER 3

BURNING OF FOSSIL FUELS

(Oil, Coal and Natural Gas)

Fossil fuels are natural energy sources found in the ground and sea depth. They are formed by accumulated dead and decayed plants and animals under intense pressure several million of years ago. They are fuels because they release heat and they are used in operating different engines and machines. They exist in three different forms. These are crude oil used in the production of petroleum products, coal and natural gas.

Crude oil is the most used among the fossil fuels. This is because it generates the greatest number of products that can be used in almost every industrial sector of a country. It is also useful for so many domestic purposes either as gas, liquid or solid form.

Reasons for Mining Fossil Fuels:

The major reasons for mining fossil fuels are; to provide energy (gas, liquid and solid forms) for operation of industries, homes, machines, automobiles and other energy dependent devices globally. Besides, fossil fuels are mined to provide variety of raw materials for road construction and production of goods such as plastic, fertilizer, wax, chemicals and gas for domestic, business and industrial uses. They are also produced to meet electricity demands.

B. Coal Mining:

Coal is not a single substance. It is a complex mixture of different chemical elements such as carbon, hydrogen, oxygen, nitrogen and sulphur. Coal mining generates many other useful products in addition to the common lignite (brown coal), bituminous (soft coal) and anthracite (hard coal). The other products that can be derived from coal mining are ammonia, coal gas, coal tar and coke.

C. Natural Gas Mining:

Natural gas is the third type of fossil fuel. They either exist separately or in combined form with crude oil. Natural gas consist mainly methane. Other types of hydrocarbon and chemical elements can be found with methane. Some of these are ethane, butane, propane, oil, water, sulphur, and so on. If natural gas exists together with oil, they are the first products that will be obtained during fractional distillation process of crude oil. Natural gas exists as liquid gas when it combines with oil or impurities. Natural gas is very fine, clean, easy and cheap to use with less pollution effect. This gas contributes to global warming when it is released into the atmosphere and broken down into components such as carbon and hydrogen.

How Fossil Fuels Cause Global Warming:

In general, all the fossil fuel products cause greenhouse effect and these result in global warming during the extraction, crushing (in the case of coal), production or processing/refining, distribution in pipelines or by trucks, storage in tanks and combustion stages. The major gases released during these different stages are carbon dioxide and methane. Others are; oxygen, nitrogen, sulphur, etc.

For example, when fossil fuel products are used in industries, at homes, by machines, by vehicles, by aircraft, by ships, by trains or to generate electricity; carbon dioxide, carbon monoxide and methane are emitted into the air (see pictures of some sources of greenhouse gases on figure 45). These gases build-up and cover the earth thereby preventing absorbed heat from the sun from returning to the space. The consequence of this continuous process is excessive or abnormal heat known as **greenhouse effect**. This, in turn, will eventually lead to many adverse situations namely melting of ice, flooding, drought, health problems, damage to crops, loss of lives and property, huge economic losses, etc.

Advantages/Uses of Fossil Fuels:

Fossil fuels are abundant under the ground and under the sea. They are used by automobiles, industries, trains, by heavy machines, airplanes, in offices, by power plants, by by ships, at homes and so on. Fractional distillation of crude oil produces products such as petrol, diesel oils, wax, asphalt, aviation fuels, kerosene, domestic and industrial gases, grease, lubricating oils, variety of chemicals for research and experiment, etc.

Disadvantages of Fossil Fuels:

Despite the high global demand for fossil fuels, it has many disadvantages that need to be considered. These happen when mining, refining and using it. Some of these disadvantages are pollution of water, soil and air. It exposes land to erosion. It is a dangerous task. It is non-renewable. It can cause serious health hazards for humans and animals. It contributes greatly to global warming. It can affect agricultural production and so on.

Control Measures: All companies must use very efficient machinery to reduce and reuse wastes. They must fit catalytic converter to their chimneys and the exhaust pipes of all their equipment. They must set targets/plans for emission reduction and recycle waste materials, gases and chemicals to make useful products.

CHAPTER 4

OZONE LAYER DEPLETING COMPOUNDS

The ozone layer is located between the stratosphere above and troposphere below. Ozone layer protects the earth thereby reducing the harmful effects of sunlight rays on the entire living and non-living organisms residing inside it. The protection function of the ozone layer (gas) is performed by absorbing much of the ultra violet radiation from the sun, reducing the intensity and the amount of the ultra violet rays that will get to the earth.

The chemical content of the ozone layer is Oxygen3 (symbol is O_3). Human activities through industrialization continue to adversely affect this life-saving gas by their direct and indirect use and release of some industrial compounds into the atmosphere. Some of these ozone depleting compounds are ChloroFluoroCarbon (CFC), SulphurHexaFluorides (SF_6), HydroFluoroEthers (HFE) and HydroFluoroCarbon (HFC). These compounds are called halogen carbons. Halogens are chemically reactive elements. These chemicals include but not limited to fluorine, chlorine, bromine and iodine. Fluorine and chlorine are the most reactive among these chemical elements.

Halogen carbons are used as refrigerants in the production of refrigerators, in air conditioners, in spray cans, as aerosol propellants, in foam manufacturing, in plastic materials, in fire extinguishers and other products.

Effects of Ozone Layer Reduction:

These compounds have resulted in reducing the thickness of the ozone layer thereby causing different health problems for human beings. Some of the health problems include but not limited to skin cancer, blindness, sunburn, suppression of immune system, early ageing of skin. Land and aquatic animals are also affected adversely by the harmful sunlight rays as this will cause great disruption in their food chains. Agricultural crops, vegetation, and livestock will be adversely affected too.

How halogen compounds are released into the atmosphere

For example, chlorofluorocarbon (CFC) is leaked from the products during manufacturing, during its use and when the products containing the compound are repaired. Leakage can also occur if the used products are not properly disposed of.

When the halogen compound (CFC) gets into the atmosphere, it undergoes a chemical reaction. This creates holes through which ultraviolet rays penetrate and separate the chlorine from CFC. Finally, the chlorine reduces ozone to oxygen.

Prevention/Control Measures:

The emission of these ozone depleting compounds into the atmosphere can be reduced by using products that are certified free of these compounds, by depending less on the use of refrigerators and air conditioners. Consumers or users of the products containing halogen hydrocarbon must follow proper disposal methods. Government must also ensure that companies comply with laws to use alternative compounds in their production processes and also apply punitive measures for disobedient ones.

CHAPTER 5

RUMINANT ANIMALS AND GLOBAL WARMING

Domestic and wild ruminant animals play a major role in contributing to greenhouse effect which results in global warming. Ruminants are mammals that feed on plants and grasses and later regurgitate the cud to re-chew it before digestion can take place. Some examples of ruminant animals are cow, sheep, goat, deer, buffalo, bison, eland, giraffe, gnu, moose, mule and so on.

The Process of Methane Gas Production in Ruminants:

Ruminants have four chambers. The four parts are; **rumen**, **reticulum**, **omasum** and **abomasum**. During rumination, food is hurriedly chewed, mixed with saliva and moved to the first two chambers where it is separated into liquid and solid (cud). During this process, bacteria, fungi, yeast and enzymes ferment and break down cellulose into digestible forms. While resting, the cud is drawn back (regurgitated) to the mouth where it is re-chewed and moved to the Omasum where inorganic minerals are absorbed. Finally, the paste and other food components moved to the Abomasum for further absorption of nutrients and digestion. Fermentation continues when the paste and its other food components move to the small and large intestines respectively.

Methane gas is produced in large quantity in the rumen primarily by methane-producing bacteria and other micro organisms which break down the complex food substances in the green plants. These substances are further broken down to generate methane and carbon dioxide. Examples include acetate, butyrate and propionate. The methane gas is released when the animal belch or fart. The methane gas produced and released by the ruminant animals is significant because its quantity and contribution to global warming is higher than the carbon dioxide produced by the same ruminant.

The reason for this higher level of methane produced by ruminant animals is due mainly to the high dependence on the consumption of grasses and other varieties of green plants. These green plants are not only bulky enough to sustain these animals but also contain different major and minor minerals, proteins and complex carbohydrates which when broken down supply them with vital natural nutritional requirements that cannot be obtained completely from processed and inorganic foods, minerals and vitamin supplements. Besides, some of these green plants are herbs. Therefore, they supply the animal system with natural medicines thereby immunizing, preventing or curing the animals of minor and major health problems.

Prevention/Control Measures:

The prevention and control measures in this case lie primarily in the hands of animal scientists, livestock feed scientists, plant scientists and livestock farmers. These four group of professionals need to work together and come up with a better livestock feed and green plants combination.

Success in this approach will greatly reduce methane production by the ruminant animals that make up the highest producer of methane namely cows, sheep and goats.

Livestock farmers have important roles to play in reducing the production of methane by their animals. Apart from the methane produced in the rumen, animal manure also generate methane which is released directly into the atmosphere when decomposed by micro organisms. The methane in the rumen can be tapped and used to produce clean energy for the farm and the local residents. Similarly, the manure can be collected together with green plants and other organic farm wastes in a pit to make compost and generate clean energy from the methane produced. This will serve as a source of energy for farm use and also for nearby residence or community.

CHAPTER 6

TERMITES' ROLE IN GLOBAL WARMING

Termites are white insects that live in the bush or forest. Termites are social insects. They form colonies that consist of a king, a queen, soldiers and workers.

There are three types of termites depending on where they live. The three types are dry wood termites (live inside dry woods or dry trees), damp wood termites (live inside rotten moist or damp dead trees or wood) and subterranean or soil termites (live in soil tunnels or clay mound inside bush or forest).

Rural dwellers know termites and can identify the different types. But termites are recognized by most people in towns, cities and urban areas by the huge damage or destruction they cause on wooden materials in many homes. Also, they can destroy soft walls, books and clothing materials. The favorite foods of termites are cellulose objects or materials particularly wood, trees and furniture items.

How do termites contribute to global warming?

Like ruminant animals, termites are able to digest cellulose because of the presence of complex carbohydrate-digesting micro-organisms in their system. They do not have four compartment stomachs like ruminant animals, but in the process of digestion, methane gas is produced and released to the atmosphere. The methane gas produced and released per termite is small but because of their large population the quantity of methane gas becomes so much that it competes with the quantity produced by higher animals such as ruminants.

Effects of Termites Activities:

Termites play different vital roles in the forest ecosystem as one of the components of the food chain. They also help in keeping the soil structure in good shape thereby reducing soil erosion and enhancing soil aeration. They enrich the soil with organic nutrients. They are good source of insect protein and fats. However, the hazard they pose by releasing methane into the atmosphere (global warming) and the huge economic loss they cause by destroying healthy trees and expensive wooden products outweigh their benefits.

Prevention/Control Measures:

The commonest preventive measure used by foresters, farmers and environmentalists in controlling termites is by spraying pesticide on the colony. Selective type of pesticide should be used as these will control termites and other types of destructive insects only. This must be performed by professionals and cautiously to avoid destroying the ecosystem. Since they release methane into the atmosphere, fire can be introduced into the mounds. This must be done with caution and monitored to avoid bush or forest fires.

CHAPTER 7

PADDY RICE FARMING AND GLOBAL WARMING

Rice is the most common food in the world. It is the most eating of all the grain crops such as wheat, barley, oat, rye, maize, guinea corn and so on. It is the second most cultivated crop in the world. Rice can be grown anywhere as long as there is enough water supply.

Where there is insufficient water, irrigation farming can be practiced to cultivate rice. To increase the quantity of rice produced per hectare, farmers also plant rice in swampy or marshy land. Where these types of areas are not available, farmers channeled water from nearby streams, rivers, lakes or from any natural bodies of water nearby to flood a dry land for the cultivation of rice. The method whereby rice is planted in water-logged field or in a field which has shallow stagnant water or swamp is known as paddy rice farming.

How does paddy rice farming contribute to global warming?

The flooded, marshy or swampy land used for paddy rice farming create ideal habitat for micro organisms that produce methane (CH_4) gas. Rice requires carbon dioxide (CO_2) to produce large quantity of grains. In other words, high carbon dioxide in the atmosphere together with warm temperature will result in faster growth and higher yield of rice plants per hectare.

Faster rice growth in turn, will energize the micro

organisms to respire more carbon dioxide that will be combined with hydrogen to produce more methane in a non-oxygen (anaerobic) water-logged soil environment. This process shows that methane gas in paddy rice farm is produced by soil micro organisms that release carbon dioxide. If you walk in paddy rice field, there is a lot of gas bubbling out. The bulk of the gas is methane.

Prevention/Control Measures:

Paddy rice farmers have many options that can be used to reduce methane gas release into the atmosphere without reducing yield. Some of the control measures include draining the field before rice maturity, planting heat tolerant rice, using straws of previously harvested rice as fertilizer, irrigating rice farm mechanically and reducing nitrogen fertilizer application.

CHAPTER 8

LANDFILL, WASTEWATER STAGNATION AND GLOBAL WARMING

Landfill is a location or site where large amounts of waste materials are dumped or disposed of by burying them in a very large, deep hole. Landfill contains different types of wastes such as paper, nylon, wooden materials, bottles, bad or damaged electronic items, metals, rubber materials, kitchen wastes, broken or old ceramics, farm wastes, damaged toys and many other bio-degradable and non-bio-degradable toxic and non-toxic wastes.

Effects of Landfills:

How does landfill contribute to greenhouse effect which results in global warming? The degradable wastes undergo continuous breakdown or decomposition by the action of micro organisms, bacteria. Methane, the most significant gas is emitted to the atmosphere during the process of microbial (mainly bacteria) decomposition of the wastes or organic matter. Carbon dioxide is also emitted to the atmosphere during this process.

Besides the emission of greenhouse gases from landfill sites, they also pose serious health problems to people. For example, flies and cockroaches from the refuse dump can transmit disease-causing micro organisms to kitchen utensils, or directly to drinks, water and food items causing dysentery, diarrhea, vomiting and other types of sicknesses primarily for pedestrians and nearby communities. It is a serious public health issue.

Prevention/Control Measures:

Consumers and industries should adopt the method of sorting, reuse and recycle of waste items. In other words, wastes should be properly sorted and placed in separate containers for recycling companies to collect, process and supplied or sold for home and industrial use. Public health and environmental officers must advice and/or enforce these waste disposal methods. Organic wastes can be used to make compost to generate energy and produce organic fertilizer. The public should be cautioned against illegal dumping of refuse and littering. Companies should be surcharged for not complying with environmental rules, regulation and procedure.

WASTEWATER STAGNATION

Stagnant dirty water is found in gutters, in pools, gullies, potholes and ditches. These contribute to global warming. Most people are not aware of this effect. Stagnant wastewater or stagnant dirty water is an ideal habitat for anaerobic micro organisms. In other words, microbes that can survive and produce methane gas in non-oxygen areas inhabit dirty water. The methane gas produced by these microbes is released continually to the air and this increases greenhouse effect and global warming. Although waste water can be seen in many locations, dirty stagnant water is common in the slums.

Causes of Water Stagnation:

There are many reasons for water stagnation both in urban and rural areas. Some of the reasons include but not limited to bad and/or poorly maintained or insufficient infrastructure such as good drainage systems to serve growing population, lack of road maintenance, lack of or violation of urban and rural development laws, poverty in slums and rural areas, lack of waste water disposal facility in buildings, lack of awareness and a host of other reasons.

Prevention of Water Stagnation:

The public should be enlightened on the adverse effects of water stagnation and the benefits of decent wastewater disposal methods. Infrastructure should be maintained regularly and new ones should be developed by government to serve the growing population. Any person who violates urban and rural development laws should be punished.

CHAPTER 9

NATURAL CAUSES OF GLOBAL WARMING

Natural causes of global warming belong to another category of global warming. These natural causes are beyond the control of human beings and they are unavoidable. Some of the natural causes are volcanic eruption, forest fires or bush fires, permafrost, earth surface, changes in sun's intensity, cloud cover, death of human and animal and decay of organic matter.

1. VOLCANIC ERUPTION AND GLOBAL WARMING

Volcanic eruption occurs in a volcano. A volcano is a hill or mountain with a hole or a channel (leading to an opening in the surface of the earth) through which magma, ash, hot gases and rock fragments escape to the surface. Any conical hill or mountain which undergoes this process is a volcano. This is one of the ways in which the earth cools off and releases internal heat and pressure.

The Causes and the Process of Volcanic Eruption:

Hills, rocks and mountains are huge, natural solid objects above the surrounding soil level. The internal condition (active, dormant, gaseous, liquid or solid) of these natural objects depends on the temperature, pressure and the chemical and/or organic matter composition. In a volcano, very high temperature and pressure cause the mantle to melt and form a complex viscous substance called magma (molten rock).

Magma contains different substances such as water, carbon dioxide, sulphur dioxide and so on. The magma is less dense or lighter in weight than the overlying rock. Therefore, intense pressure from above forced it upward. As it rises up towards the crust, the gases in the magma is released in the form of bubbles, burst open and mixed with atmospheric air (causing natural global warming). When this gas bubble is above seventy per cent, the magma bursts or explodes through the opening in the earth crust. This process is called volcanic eruption. Inflow of fresh or new magma into an active magma chamber can also cause eruption.

Necessary Measures:

Volcanic eruption cannot be prevented or stopped. Human beings only need to heed warnings or alert and respond promptly to avoid loss of life and property. To avoid loss of life and property, laws can be made to disallow people from residing in the vicinity of volcanoes.

Because the area covered by lava flow is known to be very fertile for crop production (e.g. coffee in Hawaii's Mauna Loa volcanic soil and tomato in Italy's Mount Vesuvius volcanic soil), many people stayed and farm in the vicinity of the volcanoes that recently erupted. In this case, government should provide alternative fertile farmland. Government should also set up geological survey office to monitor volcanic activity and alert residents of impending volcanic eruption.

2. NATURAL FOREST FIRES AND GLOBAL WARMING

Natural forest fires or bush fires can be caused by any of the following; dry lightning, volcanic eruption, extreme heat, underground coal deposit and dry organic matter.

DRY LIGHTNING AND FOREST FIRE

Thunder and lightning are common features before, during and after rain. Sometimes, there may be thunder and lightning without rain. Light flashes can also occur occasionally without thunder. These light flashes appear and disappear in extremely short periods. However, these light flashes cover different time spans. In other words, some light flashes have shorter period of time, these are called cold lightning. While others have longer period of time, these are called hot lightning. The shorter light flashes have higher voltage while the longer light flashes have lower voltage.

How does dry lightning cause Forest Fire?

Forest fire is a common occurrence during summer or dry season. Forest fire is aided by wind and intense heat or high temperature and dry organic matter.

When hot lightning strikes forest or bush and rain does not follow the dry or hot lightning, the forest or bush will naturally be set ablaze and spread as fast as the wind blows first in the direction of the wind and later cover the entire forest or bush. This is called **forest fire** or **bush fire** or **wild fire**.

Effects of Forest Fire:

The adverse effect of this type of fire is obvious. It destroys everything inside and around the forest. Only human beings and animals that are not caught unaware will escape. The heat generated by forest fire is extremely strong and this will surely release so much greenhouse gases namely carbon dioxide, carbon monoxide, nitrous oxide and other types of gases (toxic and non-toxic) to the atmosphere. These gases will build-up and contribute to natural global warming.

Necessary Measures:

Natural forest fire cannot be prevented. It can only be controlled from spreading to a large area if prompt actions are taken by the residents in the vicinity of the forest, by the National Emergency Management Agency, by the fire response agency and other relevant agencies.

CHAPTER 10

OTHER NATURAL CAUSES OF GLOBAL WARMING

1. **Earth's Surface**

The body or mass of objects on the surface of the earth is also one of the causes of natural global warming. The mass of object on the surface of the earth have two different natural shades namely light and dark or bright and dull. An example of light or bright natural object is snow. An example of dark or dull natural object is soil.

How does earth surface cause natural global warming?

Sun rays strike on many different objects on the earth surface at the same time. Some of the objects will reflect the rays back to space and the earth becomes bearably warm or cool. This is the effect of light or bright objects.

The more the naturally light or bright objects on the earth surface the more the reflection by these objects the more the condition will be conducive for human beings and animals.

On the other hand, when sun rays strike dark or dull objects or surface, the earth will absorb most of the sun rays and become abnormally warm or hot. This latter process will contribute to natural global warming.

This effect is even greater during summer or dry season. People living in the equatorial or tropical region experience this type of natural global warming more than those in the temperate region of the world. The level of heat or natural global warming in the tropical region depends in part on the total dark or dull areas in a particular state or country as a whole.

2. VARIATION IN SUN INTENSITY AND EARTH ROTATION

Variation in sun intensity together with earth rotation is a vital factor which contributes to natural global warming. The level or the intensity of sunlight on the earth surface will directly determine the heat experience by living and non-living things and the reaction of these two classes on earth. When the intensity is low on a particular region, the geographical region will experience a normal heat but when it is high, the region will experience abnormally higher heat. This is called **regional warming**. This happens when higher sun intensity did not last for a complete earth rotation to take place.

If the higher sun intensity continued while the earth completes its rotation on its axis or even last longer than the period for a complete earth rotation, then the entire earth surface will experience higher or abnormal heat. This is natural global warming. One vital point here is that this type of natural global warming rarely occurs because the earth is in continuous rotation or spinning.

Effects of Variation in Sun Intensity and Earth Rotation:

Increase in sun intensity and earth rotation will lead to abnormal generation of heat. The secondary effect of this process will depend on the region that is having increase in sun intensity. For example, the temperate regions will experience glazier or ice breakdown and melting. This will result in sudden increase in ocean, sea and river levels. This in most cases will lead to flooding in many low lands, coastal and below sea level areas. For the tropical regions, the adverse effects include but not limited to heat wave (a prolonged period of abnormal heat), heat exhaustion, hot dry skin, fatigue, drought, severe headache, death of human and animals, crop failure, damage to personal property and so on.

Necessary Measures:

Government should set up good weather forecast offices with modern equipment. People in areas that are prone to disaster should be alerted always and should be evacuated promptly when danger signs are noticed.

3.   **PERMAFROST**

Permafrost refers to frozen of subsoil at or below 0^0C for two or more years. This subsoil condition is common in high latitude, in mountainous areas and in temperate regions. For example, permafrost occurs regularly in Alaska, Canada, Greenland, Iceland, Siberia, north and south poles, north eastern china, north eastern Mongolia and many other cold regions.

The Causes and the Process of Permafrost:

Prolonged low temperature and cold atmospheric condition are responsible for permafrost. These two factors keep the topsoil in its normal state while the subsoil layers become frozen down the soil profile. The depth of subsoil layers that will be frozen is determined by the level of snow or ice formed and how long the snow or ice lasts.

During the period of snow formation, large quantities of organic matter are trapped in the subsoil and remain frozen for years. The permafrost soil remains in a frozen condition until there is a rise in temperature.

When there is sufficient increase in temperature, the permafrost soil layer will begin to thaw thereby setting the subsoil loose and trigger bacteria activity to decompose the high carbon organic matter and finally release carbon dioxide (CO_2) and methane (CH_4) into the atmosphere. As more greenhouse gases (CO_2 and CH_4) are released, more permafrost will thaw, the thawing will cause higher temperature, this, in turn will cause more emission of greenhouse gases. This cycle continues for as long as the temperature rises.

Effects of Permafrost:

When permafrost thaws, it contributes greatly to climate change by releasing greenhouse gases into the atmosphere. Large quantity of carbon is washed to the sea, rivers and ocean where sunlight aid the release of these gases into the atmosphere causing increase in level of these water bodies and flooding. Landslide is another bad effect of thawing permafrost. This may result in building damage, damage to roads, railways, airports, water and sewage pipes.

Subsidence (sinking of land and other structures), change in ecosystem, loss of life are other adverse effects of permafrost.

Necessary Measures:

Buildings should be constructed on stilts (long upright strong pieces of wood or metals). For big buildings, long insulated pilings should be used (e.g. Siberia buildings). Buildings can be constructed on bedrock or coarse, well-drained soil that has low ice content. Structures can also be insulated to prevent bad effect of permafrost thawing. Government should assist the public in the construction of permafrost-resistant buildings. Building construction companies can also be assisted to import modern equipment from manufacturers.

4. DEAD AND DECAYING ORGANIC MATTER

It is natural for human beings and animals to die either due to old age, sickness or by accident. Also, for one reason or the other; plants, trees and other vegetative matters dried after some period of time and they are decomposed by microbes or micro organisms on or inside the soil. Decomposition and decay of dead humans, animals and vegetative matters are carried out by bacteria, fungi, beetle larvae, maggots, termites and other micro organisms and small insects.

Effects of Dead and Decaying Organic Matters:

During the process of decomposition and decay, the

most important gases produced and released to the atmosphere in large quantities are carbon dioxide and methane. Although these gases have beneficial effects in carbon cycle for continuous existence of human, animal and plant life on earth, they also pose serious danger in causing greenhouse effect which eventually leads to natural global warming.

Necessary Measures:

Dead humans, animals and decaying organic matter should be properly disposed of and/or utilized in any beneficial way to reduce the emission of greenhouse gases to the air.

How do the greenhouse gases emitted or released to the air by natural, human and microbial activities cause global warming which ultimately leads to climate change?

CHAPTER 11

HOW GREENHOUSE GASES CAUSE GLOBAL WARMING

Greenhouse gases are gaseous compounds in the atmosphere that can absorb infrared (invisible radiant energy) radiation and trapped heat to produce greenhouse effect (excessive heat) on earth. The ultimate adverse effect of this is global warming. These gaseous compounds are produced and released by natural, human, animal and microbial activities.

Some of the major greenhouse gases include carbon dioxide (CO_2), methane (CH_4), nitrous oxide (N_2O) and halocarbon gases such as chlorofluorocarbon (CFC) and hydrofluorocarbon (HFC). Some of the minor greenhouse gases include but not limited to hydrogen and carbon monoxide. The level of greenhouse effect of a particular gas that will lead to global warming depends; on the concentration or the amount of gas, how long the gas can stay in the atmosphere and the potential or the ability of the gas to cause greenhouse effect. The following six pages will discuss the major greenhouse gases and how they cause global warming.

1. Carbon Dioxide (CO_2):

Carbon dioxide has longer wavelength than the sunlight ultraviolet rays. When these ultraviolet rays get to the earth surface, they are absorbed into the atmosphere. Carbon dioxide absorbs heat energy from the earth making the molecules unstable. The carbon dioxide molecules become stable when some amount of the absorbed heat is released back to the space and some amount is emitted back to the earth. The remaining heat is trapped by the carbon dioxide. Carbon dioxide can remain in the atmosphere for many years with the heat trapped in it causing greenhouse effect on earth. This process coupled with carbon dioxide released from many natural and human activities increase carbon dioxide greatly resulting in global warming.

For example, when forest or vegetation is destroyed, carbon that is stored for carbohydrate production (photosynthesis process) will be released to the air. Also destruction of the vegetation means that the carbon dioxide that should be absorbed by green plants and trees will remain and build up in the atmosphere. Also combustion of fossil fuels produces carbon which reacts with oxygen to form carbon dioxide.

$$C + O_2 => CO_2$$

*C: [carbon], O: [oxygen], CO_2: [carbon dioxide].

2. Methane (CH_4):

Methane also has longer wavelength than the sunlight ultraviolet rays. It is also capable of absorbing heat from sun even at a higher rate than carbon dioxide, and release small amount back to the space. Methane period in the atmosphere is shorter than carbon dioxide but releases greater amount of heat to the earth.

For example, when methane is released by natural,

human, animal and microbial activities, two thermal processes will occur. First, heat absorption from sun will occur and break the bond or link between the carbon and hydrogen (C and H_4) atoms to allow carbon to react or combine with oxygen. This is immediately followed by a greater heat releasing process that will result in the formation of carbon dioxide and water. The second thermal process shows that heat will be released to the earth causing greenhouse effect and resulting in global warming.

First Thermal Process: (heat is absorbed here)

$$CH_4 \quad C == H_4 \quad [C + 2H_2]$$

Second Thermal Process: (greater heat is released here)

$$[C + 2H_2] + 2O_2 => CO_2 + 2H_2O$$

*C: [carbon], H: [hydrogen], O: [oxygen], H_2O: [water].

3. Nitrous Oxide (N_2O):

This third most vital greenhouse gas is released by natural and human activities. Application of inorganic fertilizer to soil, volcanic eruption, nitrogen leaching into seas, rivers and ocean; manufacturing of acid and nylon production are some of the sources of nitrous oxide.

Like the halogen carbon compounds, nitrous oxide causes greenhouse effect by depleting the ozone gas (reducing the layer) in the atmosphere causing more ultraviolet rays from the sun to penetrate and abnormally increase the earth temperature resulting in global warming and finally climate change. Besides its greenhouse effect, inhalation of high concentration can cause dizziness, nausea and unconsciousness. It can also cause skin cancer because of its depleting action on the ozone layer. Nitrogen release to the air mixes with oxygen in the first instance to form nitrogen monoxide. Then nitrogen monoxide reduces ozone to oxygen to produce nitrous oxide. This is ozone depleting process. The general adverse effect of this process is global warming.

$$4NO + 2O_3 => 2N_2O + 4O_2 \text{ [ozone gas reduction]}$$

*NO: [nitrogen monoxide], O_3: [ozone],

*N_2O: [nitrous oxide], O_2: [oxygen].

4. Halocarbons (CFC, HFC):

These halocarbon compounds deplete ozone layer thereby allowing ultraviolet rays from the sun to reach the earth and increase the temperature abnormally. These compounds are released in small amount but they have strong heat-generating (greenhouse effect) potential or ability. This qualifies the halocarbons as one of the major causes of global warming.

For example, in chlorofluorocarbon compound, the halogen components (fluorine and chlorine) are the harmful factors. Chlorine being a more powerful reactant is released when the bond is broken. This is followed by reaction between chlorine and ozone which reduces ozone to oxygen. These chemical processes will continue for many years thereby depleting the ozone layer, allowing sun ultraviolet rays to reach the earth surface and finally contributing to global warming.

$$Cl_3FC \quad Cl_2FC == Cl \quad [Cl_2FC + Cl]$$

$$Cl + O_3 => ClO + O_2 \text{ [ozone gas reduction]}$$

*Cl_3FC: [trichlorofluoro carbon],

*Cl_2FC: [dichlorofluorocarbon],

*Cl: [chlorine], O_3: [ozone],

*ClO: [chlorine monoxide], O_2: [oxygen]

The different causes and effects discussed in section two combined together to produce what is known as **GLOBAL WARMING** (worldwide rise in temperature). Global warming, in turn, will produce many negative environmental impacts either at the same time or at different times in the world. Some of the common negative impacts of global warming include melting of ice in cold and mountaineous regions; rise in sea, river, ocean and lake levels; wildfires, floods, animal migration, animal extinction, heatwave, prolonged drought, change in ecosytem, loss of wildlife and aquatic animals, damage to infrastructure, agriculture and other industries; high acid content in water, etc.

Global warming (high temperature) is just one aspect of climate change. Before climate change can occur, some or many of the other factors of climate must undergo prolonged change to produce negative effects. Some of the other climatic factors include precipitation (rainfall, snowfall..), humidity, light, air, wind and soil. For example, some of the negative impacts of change in precipitation include excessive rainfall resulting in flood, loss of lives and property; stoppage of all activities and different forms of damages in the case of excessive snowfall. The overrall unusual prolonged change in some or many of these factors coupled with global warming will result in **CLIMATE CHANGE**. Global warming can also initiate changes in some climatic factors to cause climate change.

**It is the duty of every individual globally to use fuel/ energy/power wisely, to reduce emissions and to keep the environment clean always. Take action now and tell everyone.

www.ingramcontent.com/pod-product-compliance
Lightning Source LLC
Chambersburg PA
CBHW070416190526
45169CB00003B/1288